NEXTGEN LEARNERS PRESENTS

Think Like a Brain!™ Collection

Bella Cerebellum's Balance Ballet

Helping You Move, Balance, and Flow with Ease

BOOK 3

Think Like a Brain!™
Collection Overview

Think Like a Brain!™ is a wildly engaging, silly-smart children's book collection that invites kids to explore how their brains work from the inside out. Through playful characters, imaginative settings, and rhythm-rich storytelling, this collection teaches neuroscience, social-emotional learning (SEL), and self-understanding in ways children absorb without even realizing they are learning.

Instead of telling children how to behave or what to feel, Think Like a Brain!™ shows them why their brains respond the way they do. Each story transforms complex brain science into friendly characters and relatable moments, helping kids feel capable, curious, and empowered rather than confused or ashamed.

Designed for children first and embraced by parents and educators as preventative education, this collection supports:

- Emotional regulation and self-awareness
- Executive functioning skills
- Neurodivergent understanding and inclusion
- Brain-body connection and growth
- Compassion for self and others

With humor, heart, and factual accuracy at its core, Think Like a Brain!™ helps children build the skills many adults wish they had learned sooner, one playful page at a time.

One of the collection's most powerful features is how the characters become real-life reminders for children long after the story ends. Once kids learn who the characters are and what jobs they do, they can begin to recognize those same processes in themselves. A big feeling might become, "It looks like Amy is trying to tell you something," or a moment of pause might sound like, "What does your Fred think?" By naming experiences through familiar characters, children learn to notice what is happening inside their brains, reflect on it, and move forward more quickly with understanding rather than frustration.

Series 1:

The Brainy Bunch (Brain Anatomy)

Series Overview: Meet the brain's core parts as a friendly team. Kids learn what each part does and how they work together to help thinking, feeling, moving, and learning.

1. **The Brainy Bunch**
   Focus: Brain overview and teamwork
2. **Brainstem Benny & The Automatic Adventure**
   Focus: Breathing, heartbeat, survival basics
3. **Bella Cerebellum's Balance Ballet**
   Focus: Balance, coordination, practice
4. **Olly Occipital & The Lookout Tower**
   Focus: Vision, visual processing
5. **Tammy Temporal & The Soundy Sound Storm**
   Focus: Hearing, language, sound processing
6. **Perry the Parietal & The Body Map Bash**
   Focus: Touch, space, body awareness
7. **Hippocampus Hal & The Memory Maze**
   Focus: Memory, learning, recall
8. **Amy the Amygdala & The Overactive Alarm**
   Focus: Emotional alarm system
9. **Luna Lefty & Roger Righty's Mixed Up Masterpiece**
   Focus: Left/right hemispheres, teamwork
10. **Callie the Connector & The Brain Bridge**
    Focus: Corpus callosum, brain-to-brain communication

Series 2:
Cortex Castle (Frontal Lobe Executive Functions)

Series Overview: A lively castle of executive-function helpers. Kids practice planning, starting, stopping, focusing, flexible thinking, time skills, and self-talk, with Frontal Fred guiding thoughtful choices.

1. **Cortex Castle**
   Focus: Executive function overview
2. **Frontal Fred & The Decision Dilemma**
   Focus: Decision-making, prioritizing
3. **Impulsive Ian & The Button Box**
   Focus: Impulse control, pause
4. **Memory Molly & The Wandering Thoughts**
   Focus: Working memory, holding steps
5. **Focus Frannie & The Great Distraction Chase**
   Focus: Attention, refocusing
6. **Planella & The Checklist Parade**
   Focus: Planning, sequencing
7. **Flexi & The Changey-Changeroo Day**
   Focus: Cognitive flexibility
8. **Self-Talk Sally & The Thoughtful Echoes**
   Focus: Self-talk, emotional regulation
9. **Starter Stacy & Finisher Finn**
   Focus: Task initiation, task completion
10. Mi**lo the Minute Keeper & Mae the Moment Seeker**
    Focus: Time awareness, pacing
11. **The Power of Pausing with Penny**
    Focus: Pause skills, regulation

Series 3:
Busy Brain Café (Neurodivergent Affirmations)

Series Overview: A welcoming café where different brain styles are celebrated. These stories affirm neurodivergent experiences while teaching self-understanding, communication, and compassion for others.

1. **Nathan Neuron & The Bouncy Brain**
   Focus: ADHD metaphor, movement, attention
2. **Sara Sprinkle the Sensa & The Super Sensitive Day**
   Focus: Sensory sensitivity, overload tools
3. **Overwhelmed Owen & The Senses Overload**
   Focus: Sensory overload, coping strategies
4. **Interoceptive Ivy & The Inside Signals**
   Focus: Interoception, body signals
5. **Zane's Zigzag Thoughts & The Path to Peace**
   Focus: Anxiety, grounding
6. **The Busy Brain Café**
   Focus: Neurodivergent inclusion, different processing styles
7. **The Listening Light**
   Focus: Nonverbal communication, understanding
8. **The Language of Love**
   Focus: Connection, communication, compassion
9. **Trying Theo & The How-To Hiccup**
   Focus: Persistence, frustration tolerance
10. **Drowsy Drew & The Nighttime Repair**
    Focus: Sleep, brain repair
11. **Burned-Out Bonnie & The Rebuild**
    Focus: Burnout, rest, recovery

Series 4:
Steady Systems Squad (Regulation & Growth)

Series Overview: A squad that explains how brains regulate, reset, rest, and grow stronger over time, including stress waves, sleep crews, sensory filtering, and rebuilding after hard days.

1. **The Steady Systems Squad**
   Focus: Regulation systems overview
2. **Timothy Thalamus & The Traffic Tower**
   Focus: Sensory filtering, attention routing
3. **Danny Dimmer & Wake-Up Wonda**
   Focus: Arousal and alertness
4. **The Start, Stop, and Switch Station**
   Focus: Initiation, inhibition, shifting
5. **The Night Shift & The Sleepy Crew**
   Focus: Sleep systems, restoration
6. **The Stress Wave & The Return to Calm**
   Focus: Stress response, recovery
7. **The Brain Growth Gathering**
   Focus: Brain growth and development
8. **Paul's Practice Path**
   Focus: Neuroplasticity
9. **Daisy's Daydream Network**
   Focus: Daydream network, imagination
10. **Remy's Rebuild Crew**
    Focus: Recovery and healing

NEXTGEN LEARNERS PRESENTS

Think Like a Brain!™ Collection

Bella Cerebellum's Balance Ballet

Helping You Move, Balance, and Flow with Ease

BOOK 3

WRITTEN &
ILLUSTRATED BY:

MAISEL MCLAULA

NEXTGEN LEARNERS PRESS™
CINCINNATI, OHIO

This book is published by NextGen Learners Press, an imprint of NextGen Learners LLC.

Maisel McLaula is a pen name used by the author of this work.

First Edition: 2026
Printed in the United States of America
ISBN: 978-1-972860-02-1

This book is intended for educational and informational purposes only.
It is not intended to replace professional medical, psychological, or therapeutic advice, diagnosis, or treatment.
Always seek the advice of a qualified professional with any questions regarding a medical or mental health condition.

Artwork and visual elements in this book were created using a combination of original artwork, with enhanced original direction, digital illustration tools, and AI-assisted creative technologies.

A Note for My Readers:

Hi friend,
Your brain is incredible. It thinks, feels, remembers, solves, dances, creates, and so much more!
This series is here to help you understand how your brain works, and to cheer you on as you grow your own thinking superpowers.

Every part of your brain has a job.
Sometimes they get mixed up. That's okay!

When you learn what each part does, you can work with your brain instead of feeling like it's working against you.
So read, giggle, learn, and maybe even draw your own Brainy Bunch when you're done.

I'm cheering you on with every rhyme.

- Maisel McLaula

THINK LIKE A BRAIN!™

BRAIN ANATOMY COLLECTION

EXPLORE THE CASTLE INSIDE YOUR HEAD! MEET EACH ROOM AND ITS BRILLIANT RESIDENT, FROM FRONTAL FRED TO BENNY THE BRAINSTEM. WHIMSY, WONDER, AND NEURONS AWAIT!

For Juliette,
Who loves music, dancing, and joyful movement.
Even while learning how to move and groove without falling,
you laugh, you try again,
and you remind us that joy is part of the dance.

And...

To the beautifully wired, the curiously wired, and the wonderfully neurodivergent minds, may practice feel kinder than pressure.

You are exactly what a great brain looks like.

-Maisel McLaula

In the castle's west wing where movement takes flight,
Lived Bella the ballerina, graceful and light.
She danced on the railings, she spun through the halls,
With balance and rhythm and perfect footfalls.

Her job was to steady, to balance, to sway,
To help you sit up or leap into play.

She guided your muscles with rhythm and care,
From tiptoe to tumble to climbing a stair.

But one swirling day, when the weather turned wild,
Bella spun round and bumped into files!

She slipped on her toe and slid into Fred,
Who juggled five clipboards and bonked his own head!

"I don't understand!" Bella gasped as she fell,
"My balance is broken! I'm not doing well!"

Her steps were all sloppy, her spins made her dizzy,
Her pirouette power had turned into a frenzy!

Fred helped her sit down and gave her a fan,
He said, “Even balance needs its own plan.

Let’s slow it all down and reset your beat,
From the top of your head to the tips of your feet.”

They practiced with ribbons and ropes in a square,
She tiptoed on lines and bounced on a chair.

They drummed a soft rhythm and stretched to the side,
And Bella's lost balance began to abide.

She traced figure eights with her hands in the air,
She balanced on pillows and tiptoed with care.

She tapped her left foot, then her right knee,
And practiced slow breathing and counted "1-2-3."

Her motions grew smoother, her footing more sure,
Her confidence rose with each practice cure.

She giggled mid-spin and landed just right,
Her balance returned like a bird taking flight!

The Brainy Bunch cheered as she turned with delight,
She jumped and she twirled in the glimmering light.

Her timing was sharp, her rhythm on cue,
Bella was back with a dancer's debut!

She taught all her friends how to ground and to sway,
How to notice their breath and stretch into play.

She showed them that balance isn't always innate,
It's something you practice, then celebrate!

Now Bella still tumbles and trips here and there,
But she listens and learns from each flight through the air.

Her cerebellum keeps rhythm in stride,
And she always comes back with her heart full of pride.

So if you feel wobbly or spin too fast,
Or tip over chairs as you zoom right past,

Slow down your breath and try Bella's way,
Balance can build in your body each day.

Bella's Balance Tools

- Practice slow motion and find steady ground
- Use music and breath to steady your sound
- Try figure eights, or bounce and sway
- Remember: balance can grow with gross motor play

What Is the Cerebellum?

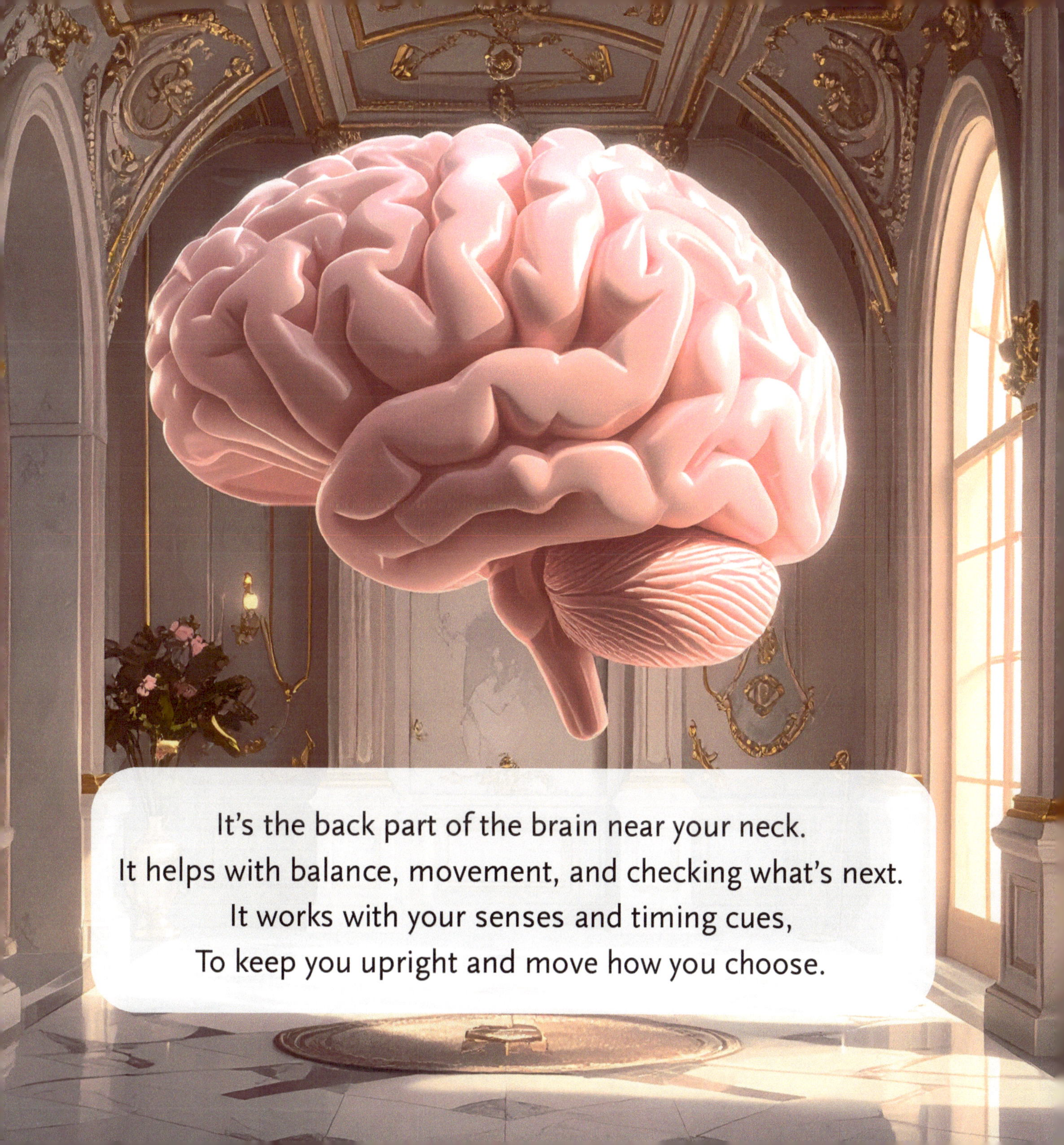

It's the back part of the brain near your neck.
It helps with balance, movement, and checking what's next.
It works with your senses and timing cues,
To keep you upright and move how you choose.

Try It Yourself!

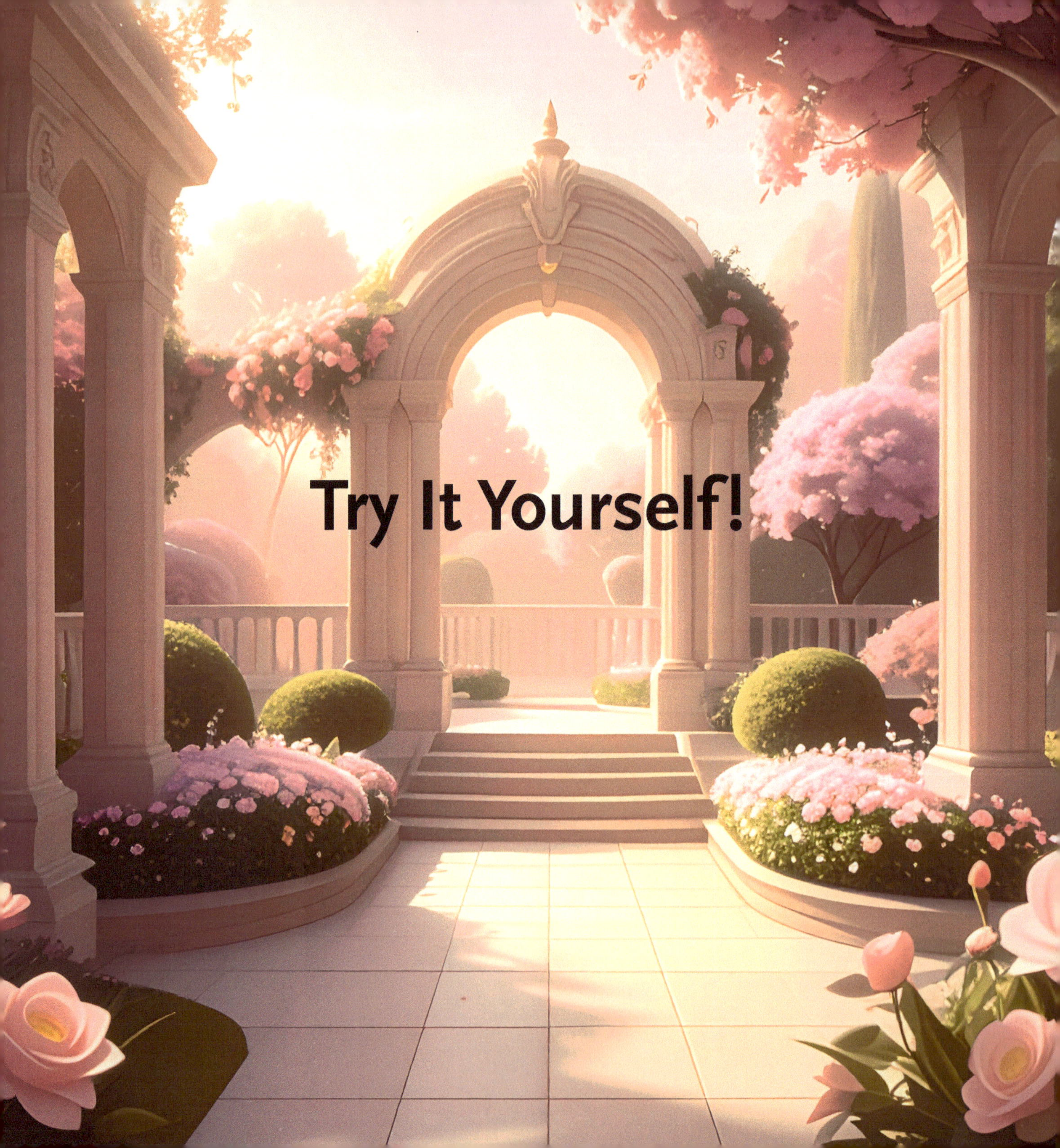

Walk on a line or balance a book on your head. Tap to a beat or try a new dance instead. Stretch one leg high or tip side to side, And see how your cerebellum helps you glide!

Congratulations!

You've danced with Bella and found your own beat,
With rhythm and motion from head down to feet.
Now you know balance isn't just luck,
It's your cerebellum helping you tuck, turn, and duck!

Thank You to My Readers

Thank you for reading this book.
Whether you're a parent, a teacher,
a therapist, or a brilliant young
brain explorer,
your interest in understanding
how the mind works
means the world.

Sharing these books
means spreading compassion and
curiosity.

You are the reason
these stories come alive.
Thank you for believing in the
power of playful learning.

Stay curious. Stay kind.
Stay brainy!

Bella Cerebellum's Balance Ballet *Teacher & Parent Guide*

💡 What This Story Teaches:

- Brain part: Cerebellum
  (movement, balance, coordination, timing)

- Core idea: Balance, coordination, and smooth movement grow through practice.

- Skills supported: gross motor control, fine motor coordination, rhythm, body awareness, motor learning

- Big message: Wobbly moments are part of learning. Practice builds pathways.

Bella shows children that movement is something the brain learns over time. Falling, wobbling, and trying again are not failures. They are how the cerebellum builds skill.

🧠 Key Concepts for Grown-Ups

Bella represents the cerebellum, located at the back of the brain near the neck. This area coordinates movement, posture, balance, timing, and motor learning. It helps actions become smoother with repetition.

The cerebellum does not learn through talking or thinking alone. It learns through doing.

When children appear clumsy, trip often, avoid physical play, struggle with handwriting, or have trouble coordinating both sides of their body, their cerebellum may still be developing these pathways.

Important foundations:

- Movement skills grow through repetition, not pressure.
- Fatigue, illness, growth spurts, or emotional stress can temporarily disrupt coordination.
- Confidence improves when effort is celebrated instead of outcome.
- Motor learning requires patience and playful practice.

This story teaches children that:

- Losing balance is normal.
- Practice changes the brain.
- Small movements build big skills.

🗣️ Discussion Starters

- What happened when Bella lost her balance?
- How did Bella get steadier again?
- What movements feel tricky for you?
- What helps your body feel strong or calm?
- Can you think of something you got better at with practice?

Encourage children to notice effort, not perfection.

👪 Try This Together:

1. Slow-Motion Walk

Walk across the room in slow motion. Notice feet, knees, hips, and arms.

2. Balance Play

Stand on one foot, then switch. Try with eyes open, then closed (with support).

3. Rhythm Moves

Clap, stomp, or march to music. Change speeds and patterns.

4. Cross-Body Reach

Touch right hand to left knee, then switch. Repeat slowly.

5. Stretch & Sway

Gentle stretches followed by side-to-side swaying help reset coordination.

🎯 Bonus Challenge

Build Bella's Balance Path

Use pillows, tape lines, or floor markers to create a simple obstacle course.

Include:

- A straight line to walk
- A soft surface to balance on
- A place to hop or spin

Ask:

- Which part was hardest?
- Which part got easier?
- How did your body feel after?

Playful Extension:

Bella Says (Like Simon Says)

Call out movement challenges:

- Bella says tiptoe
- Bella says spin
- Bella says stretch
- Bella says freeze

Make it silly. Let children take turns being "Bella."

🛠️ Bella's Balance Toolbox:

Use these supports when children struggle with coordination or movement:

- Offer daily movement breaks (jumping, crawling, dancing)
- Practice skills in short playful bursts
- Use rhythm (music, counting, clapping)
- Encourage bilateral activities (throwing balls, climbing, scooting)
- Reduce performance pressure
- Celebrate effort and improvement

Helpful phrases:

- "Your cerebellum is learning."
- "Wobbly means your brain is practicing."
- "Let's try again slowly."

❤️ Final Note for Grown-Ups

Bella reminds us that movement is brain work. Balance, coordination, and physical confidence are built through gentle repetition, playful practice, and emotional safety.

Children do not master movement by being told how. They master it by trying, wobbling, laughing, and trying again.

Every stumble teaches the brain.

Every practice builds a pathway.
Bella's ballet is not about perfection. It is about progress.

Resources & GROWTH

Resources:

(Cerebellum + balance, coordination, smooth movement, motor learning)

- National Institute of Neurological Disorders and Stroke (NINDS)
- MedlinePlus (U.S. National Library of Medicine)
- Encyclopedia Britannica
- NINDS (NIH). Brain Basics: Know Your Brain (cerebellum overview).
- Cleveland Clinic. Brain: parts and function (cerebellum/balance).
- NINDS (NIH). Brain Basics: Know Your Brain / Brain Basics resources. (2024-2025).
- Cleveland Clinic. Brain: parts and function (overview pages). (2025 update).
- CASEL. CASEL's SEL Framework (self-awareness, self-management, social awareness, relationship skills, responsible decision-making). (2020, updated).
- CDC. Child Development: Positive Parenting Tips and developmental guidance by age. (2024-2025).

Growth:

(If you or someone you know wants to improve coordination, balance, or learning through movement)

- CDC - Child Development
  - Search: motor development
- American Occupational Therapy Association (AOTA)
  - Look for: coordination, gross motor skills

Acknowledgments:

To my students: past, present, and future:
You taught me more than any textbook ever could. By watching how you explored, questioned, struggled, and discovered, I learned how minds truly grow. Every “aha” moment in these pages began with you.

To the parents, teachers, and guides walking beside children every day:
Thank you for listening, for advocating when things feel complicated, and for believing that understanding a child’s inner world truly matters. Your patience and care shape futures in ways you may never fully see.

To my husband, my best friend, and my constant partner in learning:
Thank you for walking beside me, believing in this work, and holding space for creativity to exist. Sharing curiosity, growth, and creation with you has been one of the greatest honors of my life.

To the great thinkers who shaped how I see the world: including Dr. Seuss, Albert Einstein, Leonardo da Vinci, Stephen Hawking, Ludwig van Beethoven, just to name a few, and my English teacher, Mrs. Godfrey:
Your curiosity, courage, and imagination reminded me that learning is limitless, creativity is powerful, and questions matter. Mrs. Godfrey, your encouragement, belief in me, and love of words sparked my writing journey. If not for you, I may never have picked up a pen.

And to my publisher, NextGen Learners™:
It was by fate that we met, and by inspiration that we now move forward. Together, we build something meaningful. Together, we leave this legacy.

Lastly, to my own brain:
Complex, resilient, ever-adapting, and endlessly learning. Thank you for carrying stories, holding wonder, and reminding me that growth is always possible.

Maisel McLaula is the author of the Think Like a Brain!™ children's book collection, a story-first series that blends neuroscience, imagination, and emotional safety for young readers.

Drawing from decades of experience supporting children and families, her work centers on a simple but powerful belief: children deserve to understand and accept who they are, including how their brains work and why they make the choices they do.

Through playful characters and gentle metaphors, Maisel's stories help children discover the "why" behind their thoughts, feelings, and actions, turning confusion into clarity and self-judgment into self-compassion. By understanding their inner systems, children gain the power to respond with kindness toward themselves and to recognize those same patterns in others.

Maisel believes that when children feel safe inside their own minds, empathy grows naturally. The Think Like a Brain!™ collection offers a foundation for confidence, understanding, and care that grows with the child.

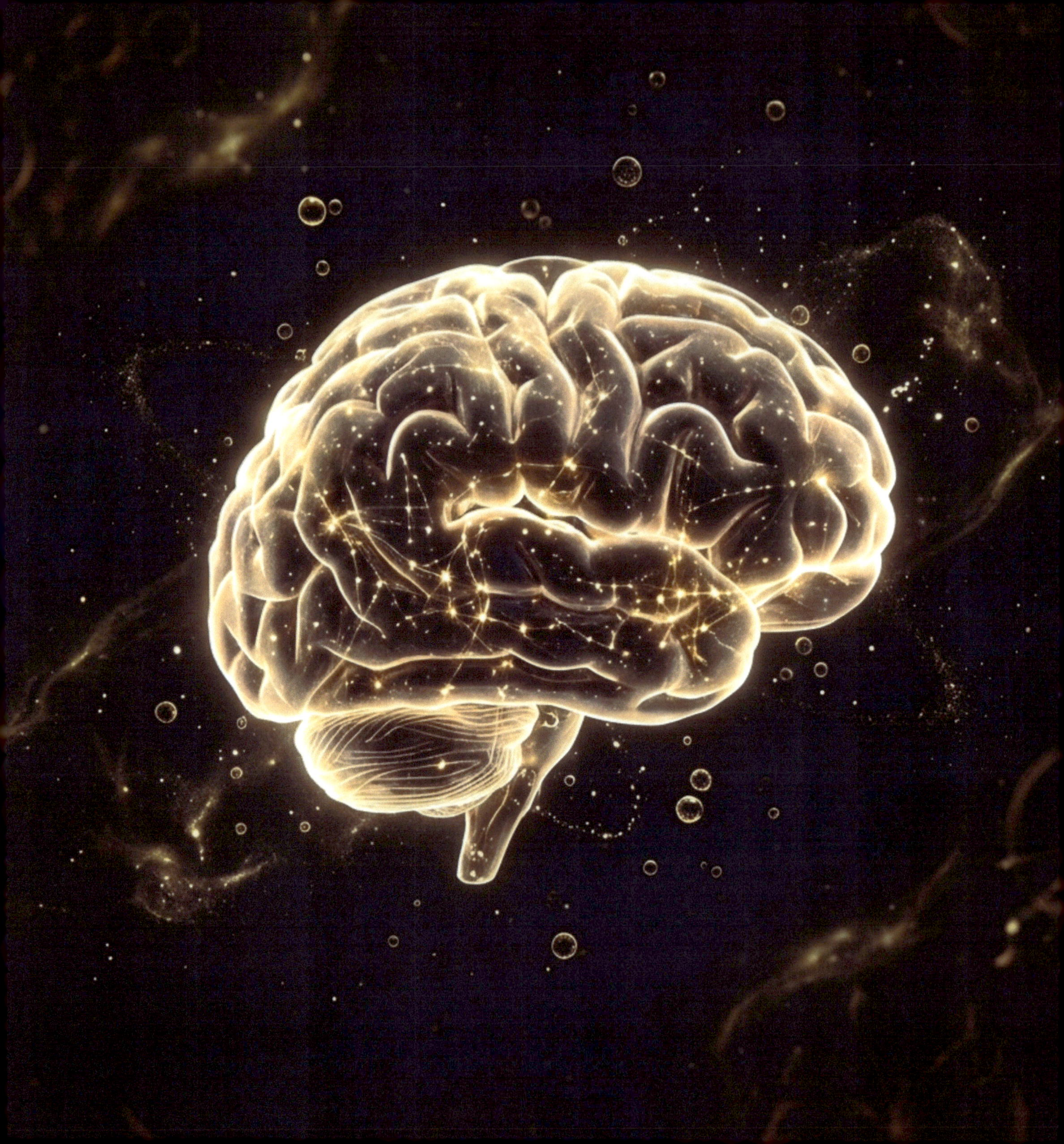

ABOUT THE PUBLISHER

NextGen Learners LLC is an educational publishing company dedicated to helping children understand how they think, feel, learn, and grow.

Through imaginative storytelling, neuroscience-informed concepts, and real-world application, NextGen Learners™ creates books and resources that support emotional development, critical thinking, self-awareness, and confidence in young minds.

Our work is designed not only for children, but for the parents, educators, caregivers, and professionals who walk beside them. Each book serves as a bridge, helping adults better understand the child in front of them, while helping children better understand themselves.

By creating shared language and meaningful connection points, our stories foster deeper relationships, reduce misunderstanding, and encourage acceptance, both within oneself and toward others.

We believe every child's mind is uniquely wired, and that understanding those differences is the foundation for compassion, growth, and lifelong learning.
With multiple collections designed to support cognitive, emotional, and developmental growth, NextGen Learners provides tools that children can carry with them, and tools that adults can use to guide them with clarity, patience, and purpose.

NG NEXTGEN LEARNERS PRESS

MORE TO EXPLORE

The Brainy Bunch is just the beginning.
More stories, tools, and brain-friendly resources are on the way! Designed to help children, parents, and educators better understand how the brain works in everyday life.

FREE PRINTABLES

We've created printable pages to go along with this book, including:

-Reading Guides
-Brain helper visuals
-Character Cards

These are available for free as part of our growing resource library.

Our full website experience is built to include:

-Downloadable activity packs
-Parent & teacher guides
-Expanded learning tools
-Additional collections and topics

STAY CONNECTED

We love hearing how children connect with these stories!

If there is a topic, feeling, or challenge you would like to see explored in future books, we would love to hear from you!
Contact us: Inspire@NextGenLearners.com
NextGenLearners.com

Your ideas help shape future stories and support the growth of this collection.

This collection grows with you. New resources and stories are added regularly.

Maisel McLaula loves hearing from the tiny humans who read her books.

Did a story make you giggle, spark a big idea, or inspire you to draw your very own character? She would be delighted to see it! Send your letters and artwork to:

NextGen Learners
Attn: Maisel McLaula
8666 Beechmont Ave #1063
Cincinnati, OH 45255

Every letter is read with a smile, and your artwork just might make her whole day.

OPEN BRAIN

www.ingramcontent.com/pod-product-compliance
Lightning Source LLC
Chambersburg PA
CBHW041322210926
51847CB00002B/23
9781972860021